MW01223261

JUMP

JOURNAL FOR UNDERSTANDING MATHEMATICAL PRINCIPLES

Eliza Akana
Jonelle Flight
Suzanne Forbes
Ann Watanabe

Common Core Education, Inc.
Maui, Hawaii

No part of this publication may be reproduced, stored in a retrieval system, or transmitted in any form or by any means, electronic, mechanical, photocopying, recording, or otherwise, without the written permission of the publisher. For information regarding permission, write to Common Core Education, Inc., 1695 Olinda Road, Makawao, Maui, HI 96768

ISBN-13: 978-0615454375

ISBN-10: 0615454372

Text copyright © 2011 by Common Core Education, Inc. All rights reserved.

Printed in the United States of America

Common Core State Standards for Mathematics © Copyright 2010. National Governors Association Center for Best Practices and Council of Chief State School Officers.
All rights reserved.

Cover design by Jessica Matsumoto

This **JUMP** belongs to:

_____

Teacher:

_____

Grade:

_____

School:

_____

© Common Core Education, Inc.

Signal Words

Signal words are important words that tell you what you need to do on each journal page.

Draw	Draw a picture or a model.	Draw
Label	Give a name to something you have drawn or written.	Label
Write	Write words or numbers.	Write
Solve	Find the solution or answer.	Solve
Explain	Put your thinking into complete sentences.	Explain

© Common Core Education, Inc.

Write a multiplication equation.   **Write** a word problem for your multiplication equation.   **Solve** your word problem.

Write

Write

_____

_____

_____

_____

_____

_____

_____

_____

_____

Solve

© Common Core Education, Inc.

Write a division equation. **Write** a word problem for your division equation. **Solve** your word problem.

Write

Write

_____

_____

_____

_____

_____

_____

_____

_____

_____

Solve

CCSS.3.OA.2

© Common Core Education, Inc.

Jean had 24 inches of ribbon that she cut into six equal pieces. How long is each piece of ribbon? Draw a picture, write an equation using a symbol to represent the unknown number, and solve this word problem.

Draw

Write

Solve

© Common Core Education, Inc.

Devon bought seven boxes of light bulbs. Each box contains eight light bulbs. How many light bulbs did Devon buy in all? Draw a picture, write an equation using a symbol for the unknown number, and solve this word problem.

Draw

Write

Solve

CCSS.3.OA.3—B

© Common Core Education, Inc.

Mr. Wu is having a party. There will be a total of 15 people at the party. There are three party hats in a package. How many packages of party hats will Mr. Wu need so that each person gets one hat? Draw a picture, write an equation using a symbol for the unknown number, and solve this word problem.

Draw

Write

Solve

© Common Core Education, Inc.

There are six rows of orange trees with nine orange trees in each row. How many orange trees are there? Draw a picture, write an equation using a symbol for the unknown number, and solve this word problem.

Draw

Write

Solve

© Common Core Education, Inc.

Mrs. Ruiz has 28 student desks arranged in four rows. How many student desks are in each row? Draw a picture, write an equation using a symbol for the unknown number, and solve this word problem.

Draw

Write

Solve

© Common Core Education, Inc.

A movie theater has a total of 100 seats arranged in a square. There are an equal number of rows and seats in each column. **Draw** an array to show how the seats are arranged. **Write** an equation for your array.

Draw

Write

CCSS.3.OA.3—F

© Common Core Education, Inc.

9 x ? = 63

Solve for the unknown number. **Explain** the strategy you used to solve for the unknown number.

Solve

Explain

_____

_____

_____

_____

6 = ? ÷ 7

Solve for the unknown number. **Explain** the strategy you used to solve for the unknown number.

Solve

Explain

_____

_____

_____

_____

© Common Core Education, Inc.

Draw or write an example of the associative property of multiplication. Explain whether or not this property would work for division.

Draw or Write

Explain

_____

_____

_____

_____

_____

_____

_____

_____

CCSS.3.OA.5—A

© Common Core Education, Inc.

Draw or write an example of the commutative property of multiplication. Explain whether or not this property would work for division.

Draw or Write

Explain

_____

_____

_____

_____

_____

_____

_____

_____

© Common Core Education, Inc.

Draw or write an example of the distributive property of multiplication. Explain how this property can help you multiply large numbers.

Draw or Write

Explain

_____

_____

_____

_____

_____

_____

_____

_____

CCSS.3.OA.5—C

© Common Core Education, Inc.

56 ÷ 7 = ?

Explain how you could use multiplication to solve this division equation.

Explain

_____

_____

_____

_____

_____

_____

_____

_____

_____

_____

_____

_____

_____

_____

© Common Core Education, Inc.

Write all of the strategies you have learned that help students memorize their multiplication facts. **Explain** which of these strategies works best for you.

Write

_____
_____
_____
_____
_____
_____
_____
_____

Explain

_____
_____
_____
_____
_____
_____
_____
_____

CCSS.3.OA.7

© Common Core Education, Inc.

Matt went to the farmer's market and bought five bags of apples with six apples in each bag. He ate four apples on his way home. How many apples does he have left? **Write** an equation using *a* to represent the unknown. **Solve** the equation. **Explain** step-by-step how you solved the equation.

Write

Solve

Explain

_____

_____

_____

_____

_____

_____

_____

_____

_____

_____

© Common Core Education, Inc.

Jimmy has five boxes of books. There are nine books in each box. Jimmy wants to donate 47 books to the library and believes he will have two books left over to give to his teacher. Think about the problem and use mental computation. **Explain** whether or not it is reasonable for Jimmy to believe he will have two books left over to give to his teacher.

Explain

_____

_____

_____

_____

_____

_____

_____

_____

_____

_____

_____

_____

_____

_____

_____

© Common Core Education, Inc.

	1	2	3	4	5	6	7	8	9	10
1	1	2	3	4	5	6	7	8	9	10
2	2	4	6	8	10	12	14	16	18	20
3	3	6	9	12	15	18	21	24	27	30
4	4	8	12	16	20	24	28	32	36	40
5	5	10	15	20	25	30	35	40	45	50
6	6	12	18	24	30	36	42	48	54	60
7	7	14	21	28	35	42	49	56	63	70
8	8	16	24	32	40	48	56	64	72	80
9	9	18	27	36	45	54	63	72	81	90
10	10	20	30	40	50	60	70	80	90	100

What pattern do you see in the multiplication table that represents the commutative property of multiplication? Explain the pattern.

Explain

_____

_____

_____

_____

_____

_____

_____

© Common Core Education, Inc.

Write a three-digit number. **Explain** how you would round your number to the nearest ten. **Explain** how you would round your number to the nearest hundred.

> **Write**

> **Explain**
> _____
> _____
> _____
> _____
> _____
> _____
> _____

> **Explain**
> _____
> _____
> _____
> _____
> _____
> _____
> _____

CCSS.3.NBT.1

© Common Core Education, Inc.

Write a two-digit number and a three-digit number. **Solve** for the difference. **Explain** which operation you can use to check your answer.

Write

Solve

Explain

_____

_____

_____

_____

_____

_____

_____

_____

© Common Core Education, Inc.

Solve 5 x 40 using the distributive property and multiples of ten. **Explain** your answer.

Solve

Explain

_____

_____

_____

_____

_____

_____

_____

_____

_____

_____

© Common Core Education, Inc.

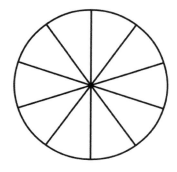

Tom ate three pieces of this pie. **Write** the fraction that represents the total number of pieces he ate. **Explain** what each part of your fraction means.

Write

Explain

_____

_____

_____

_____

_____

_____

_____

_____

_____

© Common Core Education, Inc.

The number line below equals one whole. Draw points to partition the interval 0 to 1 into eight equal parts. Label the point that represents $\frac{1}{8}$

Explain how you know it is $\frac{1}{8}$

Draw and Label

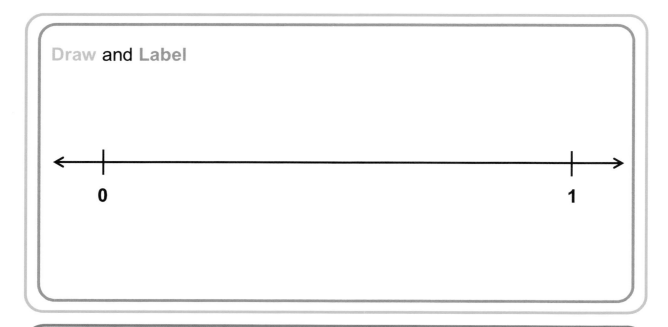

0   1

Explain

_____

_____

_____

_____

_____

_____

_____

_____

© Common Core Education, Inc.

The number line below represents one whole. Draw points on the number line. Label the point that represents $\frac{3}{4}$ Write the value of each equal part in this interval. Explain how you know.

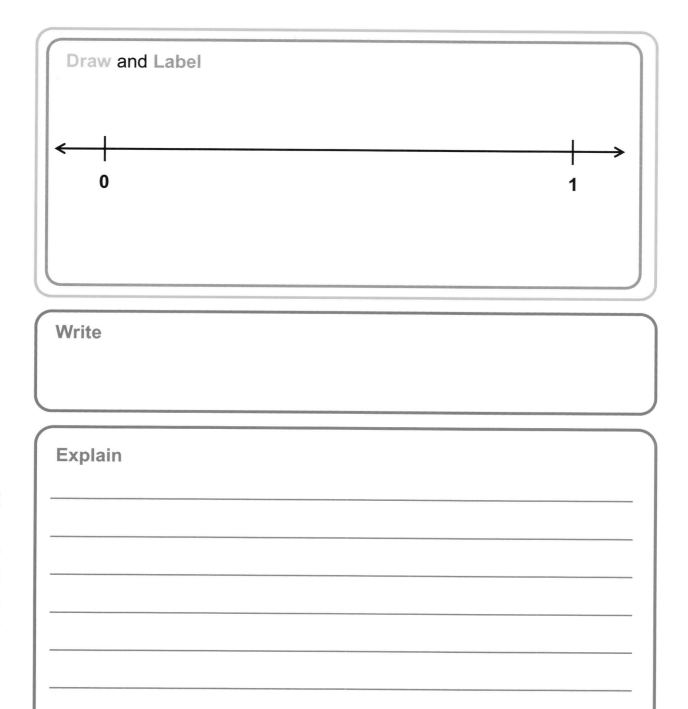

Draw and Label

|←———|——————————————————————|—→
       0   **1**

Write

Explain

_____

_____

_____

_____

_____

_____

© Common Core Education, Inc.

Draw a model or a number line to show whether $\frac{3}{4}$ and $\frac{7}{8}$ are equivalent. **Explain** how you know.

Draw

Explain

_____

_____

_____

_____

_____

_____

_____

_____

_____

CCSS.3.NF.3.a

© Common Core Education, Inc.

Write two fractions that are equivalent. **Draw** a model or a number line to prove that they are equivalent. **Explain** how you know they are equivalent.

Write

Draw

Explain

_____

_____

_____

_____

_____

_____

© Common Core Education, Inc.

Write an equivalent fraction for the whole number 4. **Draw** a model that represents your fraction. **Explain** your model.

Write

Explain

_____

_____

_____

_____

_____

_____

CCSS.3.NF.3.c

© Common Core Education, Inc.

Explain whether or not 1/2 of a basketball is equal to 1/2 of a golf ball.

Explain

_____

_____

_____

_____

_____

_____

_____

_____

_____

_____

_____

_____

_____

_____

_____

_____

© Common Core Education, Inc.

Write $\frac{1}{2}$ and $\frac{5}{8}$ with a symbol to show which number is greater. **Draw** a model to justify your answer.

Write

Draw

CCSS.3.NF.3.d—B

© Common Core Education, Inc.

Draw the hands on the clock to show 3:12 pm.

Draw

```
        12
    11      1
  10          2
  9            3
   8          4
     7      5
        6
```

Draw the hands on the clock to show that 49 minutes have elapsed.

Draw

```
        12
    11      1
  10          2
  9            3
   8          4
     7      5
        6
```

Write a word problem using the information on the clocks.

Write

_____

_____

_____

_____

_____

_____

_____

© Common Core Education, Inc.

Lindsay is having a party and needs to purchase 18 liters of orange juice. Orange juice is packaged in 2-liter containers. How many containers of orange juice will Lindsay need to buy? Draw a picture. Write an equation. Solve the equation. Explain how you found your answer.

Draw

Write

Solve

Explain

_____

_____

_____

_____

_____

_____

CCSS.3.MD.2

© Common Core Education, Inc.

Favorite Pet	
Pets	**Tally**
Dog	JHT III
Cat	JHT JHT I
Rabbit	II
Reptile	IIII

Draw a bar graph to represent the data in the table above. For your graph, use the scale 1 square = 2 pets. Label your graph.

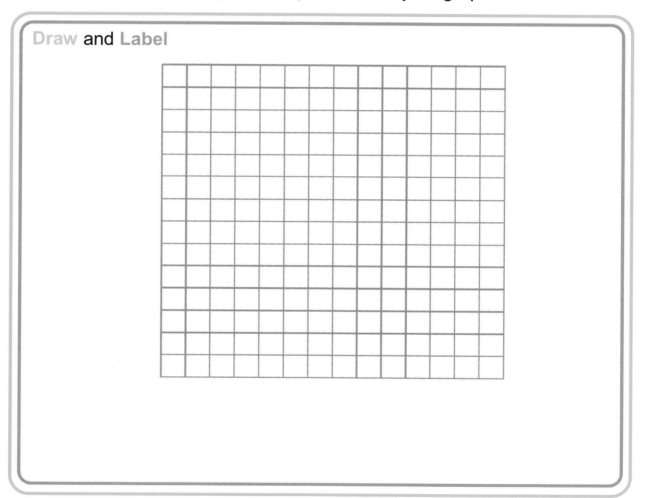

Draw and Label

Write a question about your data using the phrase "how many more?"

Write

_____

© Common Core Education, Inc.

Draw a model of a ruler that is three inches long. Label the ruler with whole numbers, halves, and quarters.

Draw and Label

CCSS.3.MD.4

© Common Core Education, Inc.

Draw three different plane figures. Each plane figure must have an area of 24 square units. Explain how you know these figures have the same area.

Draw

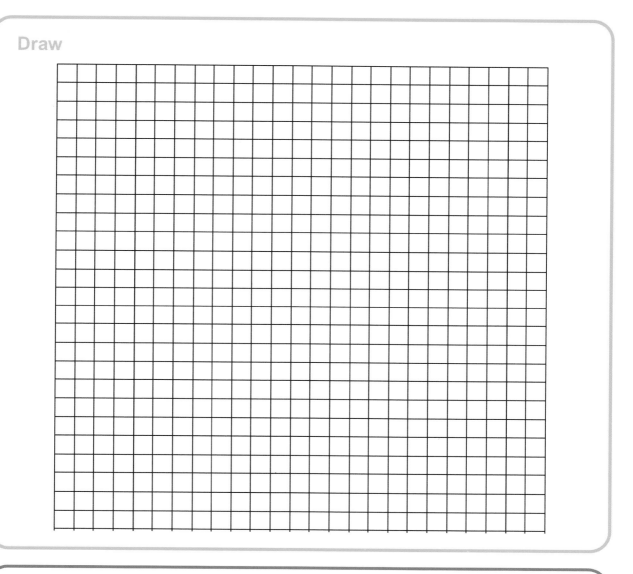

Explain

_____

_____

_____

_____

© Common Core Education, Inc.

Explain how you would measure the area of a room using square tiles.

Explain

_____

_____

_____

_____

_____

_____

_____

_____

_____

_____

_____

_____

_____

_____

_____

_____

_____

_____

© Common Core Education, Inc.

CCSS.3.MD.5.b

Write an example of when you would measure an area using square inches.

Write

_____

_____

_____

_____

_____

_____

Write an example of when you would measure an area using square yards.

Write

_____

_____

_____

_____

_____

_____

© Common Core Education, Inc.

Draw a rectangle.   Label the side lengths of the rectangle.   Write a multiplication equation for the area of your rectangle.

Draw and Label

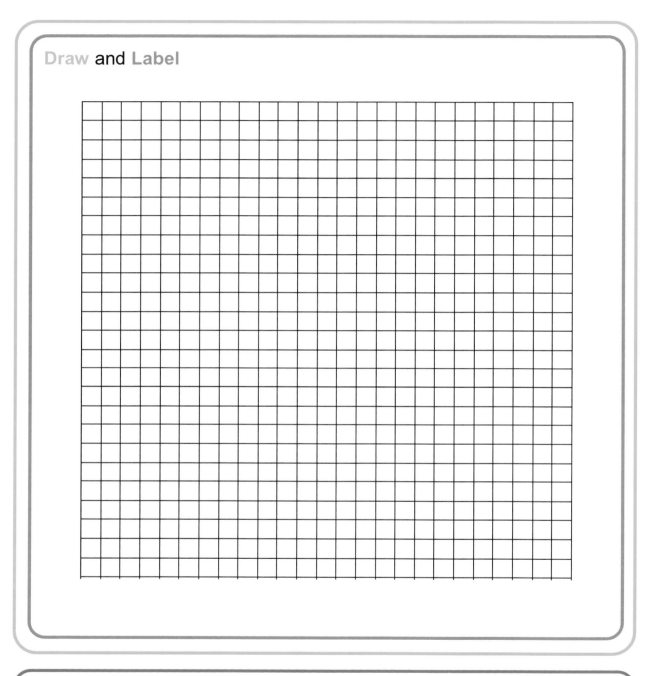

Write

CCSS.3.MD.7.a

© Common Core Education, Inc.

Write a real life word problem in which you would need to find the area. **Draw** a model to represent your word problem. **Write** an equation to represent your word problem. **Solve** your word problem.

Write

_____

_____

_____

_____

_____

Draw

Write

Solve

© Common Core Education, Inc.

Nick drew a rectangle that has an area of 32 square inches. Draw a possible example of Nick's rectangle. Label the side lengths of the rectangle. Write a multiplication equation for the area of the rectangle.

Draw and Label

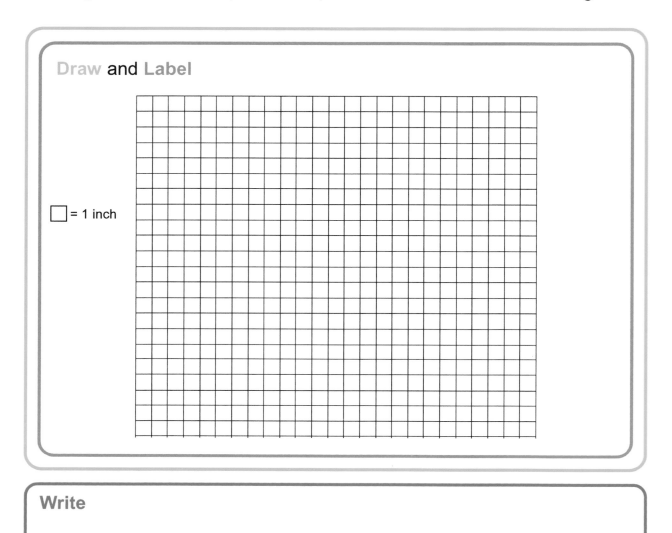

= 1 inch

Write

Nick divided his rectangle into two separate rectangles. (On the next page) Draw a possible example of the two new rectangles. Label the side lengths of the new rectangles. Write a multiplication equation for the area of each new rectangle. Add the areas together. Explain how this represents the distributive property.

(This prompt is continued on the next page.)

CCSS.3.MD.7.c (Page 1 of 2)

© Common Core Education, Inc.

Draw and Label

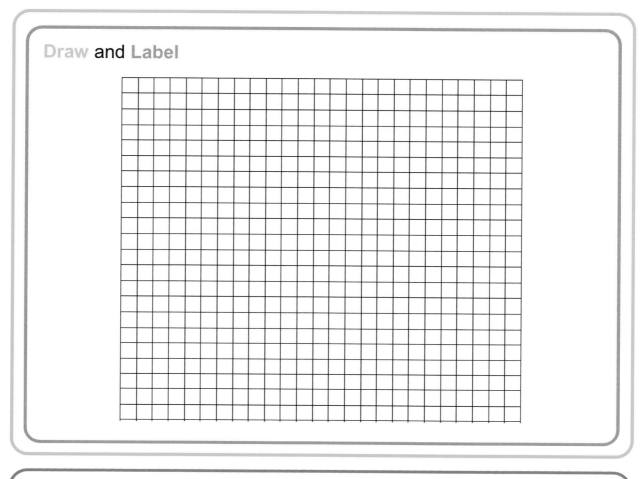

Write

Explain

_____

_____

_____

_____

© Common Core Education, Inc.

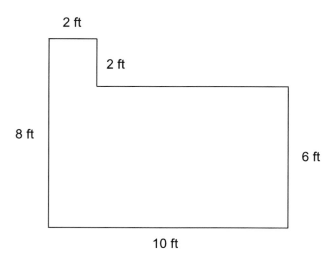

2 ft

2 ft

8 ft

6 ft

10 ft

Lucy needs to find the area of her backyard. **Solve** for the area of Lucy's backyard using the drawing above. **Explain** how you solved the problem.

Solve

Explain

_____

_____

_____

_____

CCSS.3.MD.7.d

© Common Core Education, Inc.

Draw two rectangles that have the same perimeter but different areas.

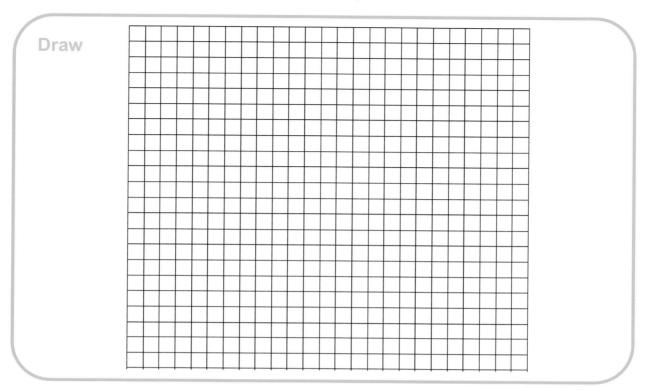

Draw two rectangles that have the same area but different perimeters.

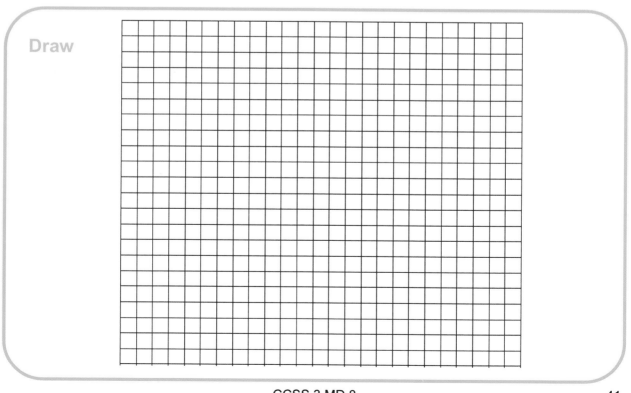

© Common Core Education, Inc.

Draw and label three different quadrilaterals. Explain how you know they are quadrilaterals.

Draw and Label

Explain

_____

_____

_____

_____

_____

_____

CCSS.3.G.1

© Common Core Education, Inc.

Draw a shape. Draw lines to partition the shape into four equal parts. Explain what fraction of the area each part represents.

Draw

Explain

_____

_____

_____

_____

_____

_____

© Common Core Education, Inc.

© Common Core Education, Inc.

Academic Vocabulary Prompts

© Common Core Education, Inc.

Multiplication

Draw a picture or model that shows **multiplication**. Label your picture or model. **Explain** what **multiplication** means.

Draw and Label

Explain

_____

_____

_____

_____

_____

_____

_____

© Common Core Education, Inc.

Division

Draw a picture or model that shows **division**. Label your picture or model. **Explain** what **division** means.

Draw and Label

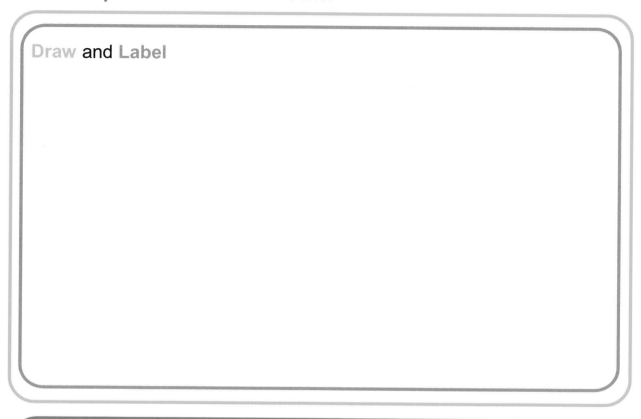

Explain

_____

_____

_____

_____

_____

_____

_____

© Common Core Education, Inc.

Equation

Write a multiplication or division **equation** using a symbol to represent an unknown number. What is an **equation**? Explain.

> **Write**

> **Explain**
>
> _____
>
> _____
>
> _____
>
> _____
>
> _____
>
> _____
>
> _____
>
> _____
>
> _____
>
> _____
>
> _____
>
> _____
>
> _____
>
> _____
>
> _____

© Common Core Education, Inc.

Strategy

Write two **strategies** that you can use to solve multiplication or division problems. **Explain** what the word **strategy** means.

Write

_____

_____

_____

_____

_____

_____

Explain

_____

_____

_____

_____

_____

_____

_____

© Common Core Education, Inc.

Fraction

Write a **fraction**. Draw a picture to represent your **fraction**. Explain the parts of a **fraction** and what they represent.

Write

Draw

Explain

3.AV.5

© Common Core Education, Inc.

Volume

Draw 3 items that have **volume**. Label your items with the item name.
What is **volume**? Explain.

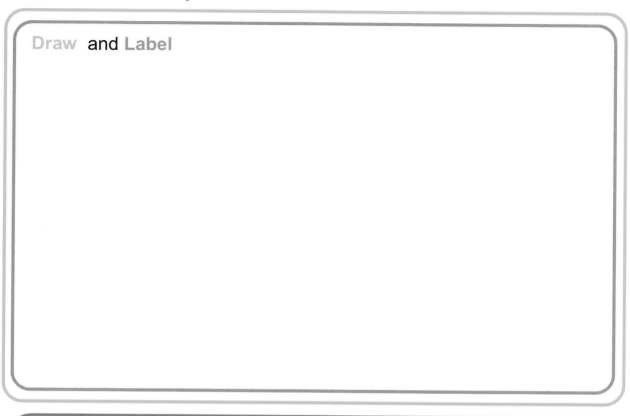

Draw and Label

Explain

_____

_____

_____

_____

_____

_____

© Common Core Education, Inc.

Mass

Draw 3 items that have **mass**. Label your items with the item name. What is **mass**? Explain.

Draw and Label

Explain

_____

_____

_____

_____

_____

_____

_____

© Common Core Education, Inc.

Data

Write an example of **data** that you could collect from your classmates. What is **data**? Explain.

Write

_____

_____

_____

_____

Explain

_____

_____

_____

_____

_____

_____

_____

_____

_____

_____

© Common Core Education, Inc.

Area

Draw an example of an **area** model or array. What is **area**? Explain.
Write two examples of when you would need to find **area** in the real world.

Draw

Explain

_____

_____

_____

_____

_____

_____

_____

Write

_____

_____

_____

_____

© Common Core Education, Inc.

Perimeter

Draw a plane figure. Label the side lengths. Solve for the **perimeter**. What is **perimeter**? Explain.

Draw and **Label**

Solve

Explain

_____

_____

_____

_____

Write

_____

_____

_____

_____

© Common Core Education, Inc.

Property

Draw and label a two-dimensional shape. Write the properties of your shape. What is a property? Explain.

Draw and Label

Write

_____

_____

_____

Explain

_____

_____

_____

_____

_____

© Common Core Education, Inc.

Classify

Draw and label a square. **Write** the different ways a square can be **classified**. What does it mean to **classify** a shape? Explain.

Draw and Label

Write

_____

_____

_____

Explain

_____

_____

_____

_____

_____

© Common Core Education, Inc.

11909727R00034

Made in the USA
Charleston, SC
29 March 2012